T0193561

QUANTUM MECHANICS: COLLAPSE

Resonant Response & Photon Annihilation

ANTONY BOURDILLON

© 2024 Antony Bourdillon. All rights reserved.

No part of this book may be reproduced, stored in a retrieval system, or transmitted by any means without the written permission of the author.

AuthorHouse™
1663 Liberty Drive
Bloomington, IN 47403
www.authorhouse.com
Phone: 833-262-8899

Because of the dynamic nature of the Internet, any web addresses or links contained in this book may have changed since publication and may no longer be valid. The views expressed in this work are solely those of the author and do not necessarily reflect the views of the publisher, and the publisher hereby disclaims any responsibility for them.

Any people depicted in stock imagery provided by Getty Images are models, and such images are being used for illustrative purposes only.
Certain stock imagery © Getty Images.

This book is printed on acid-free paper.

ISBN: 979-8-8230-1847-0 (sc)
ISBN: 979-8-8230-1848-7 (e)

And by UHRL
ISBN: 979-8-3507-2387-9

Library of Congress Control Number: 2023922713

Print information available on the last page.

Published by AuthorHouse 12/19/2023

authorHOUSE®

for Alice

Contents

Preface

This book is different from all others on the topic. We do physics. It is different from math, though you cannot do exact science without numbers. The two operate under strict disciplines. Physics is empirically falsifiable: a proposition becomes, after testing, 'logically true'. By contrast, math is thought out *a priori* from axiom to theorem. It is no more than hypothetical; never 'true'. In physics, a proposition is meaningless if it is not falsifiable; axioms are not falsifiable. They are chosen and are not substantive in the logic of physics; their existence is ephemeral, *i.e.* when math is a numerical tool. By contrast, beings must not be multiplied without necessity. This requirement is core to the credibility of physics. There are other differences that will be discussed in due course, and we will find that the rules are relaxed in computation.

How can the photon demonstrate momentum without rest mass? Why is the momentum of a wave-particle not proportional to its wavelength, as is its wave velocity? How small is a 'point particle'? What, in physics, is an 'uncertainty limit'? What physical measurement is ever made, that is instantaneous and discontinuous? How is a macroscopically dispersed wave-packet (as in Young's slits) reduced (significantly) to atomic scales during photon counting measurements? These are some of many weird features adopted without explanation in conventional quantum theory. We have, in previous papers, explored *physical* answers to all of these questions. We take them up again here to unify the interacting conclusions.

Quantum theory is not weak; it has been successful in atomic, nuclear and elementary particle physics. However there remain unanswered questions in gravity, and unease with the disunity of so many particles that are called elementary. In this treatise, we consider logical inconsistencies. The logical systems used in mathematics and physics are mature and comparatively uncontentious, and there can be no doubt that physics requires numbers, as tools for measurement: these are "exactly" required to explain and understand observed physical processes and events. How so?

'Orthodox' quantum theory is developed from wave-particle duality in photons, electrons, *etc*. The theory employs special features for minute structures. Orthodox quantun theory is partly axiomatic and partly experimental so that its logic is neither mathematical nor physical. Rather, it adapts selected features of each discipline for computational purposes. The result is not the convincing undeniability of conventional mathematics nor the empirical reality of physics; but a speculative system for predicting the future and for calculating probable distributions for experimental outcomes. The method is not physically explanatory in terms of cause and effect, but predictive by means of adding or multiplying respectively parallel or serial probabilities that are due to force constants, masses *etc*. With examples, we compare the three logical systems.

We begin by considering the individual and established logic of math and physics employed to describe the peculiar methods of computation used in quantum theory and statistical mechanics. There is inevitable confusion when possibilities or probabilities are mixed with empirical physics. Individual atoms in an ideal gas are not measured, but we know, from math, that their energies obey Boltzmann's distribution. But what reality do we give to Born's probability amplitude for electrons in atomic or solid or liquid states? Whatever is not understood, is not physics. For us, quantum reality lies in the wave-group. It is both wave and particle.

By understanding the logic of physics and math, we learn what went wrong. We can progress to eliminate the schismatic beliefs and anomalies that have grown since quanta were discovered. The greatest is perhaps the problem of condensation. The solution we offer is derived from multiple features of modern physics. They include wave-particle duality; special relativity; the wave-group; Dispersion Dynamics; Uncertainty; quasicrystal diffraction; harmonics in quanta; and resonant response. The resonance within the wave-group is the key that unlocks condensation. It is not rocket-science; nor world politics; nor an asteroid on collision path; just first principles, physical, quantum mechanics.

Generally, examples are cited from literature. However, to quickly keep track of a point that is being made, a graphical representation of calculated results is typically provided. These are convenient markers to the argument. The book supposes an introductory interest in fundamental physical theory; the purpose given for this writing is to undo unnecessary complication in the practice of physics. Here, quanta are simply explained; they are nowhere weird.

AJB, MA DPhil (Oxford) PhD (Cambridge) LicPhil (Heythrop)

San Jose, CA

1. Introduction

Condensation, or the collapse of the wave packet, has been mysterious since the early conception of quantum mechanics. If you think you know what quantum theory is, skip this introduction to search for what is new beyond it. Following wave-particle duality in 19th century physics, the *problem*—of how at absorption or measurement, the quantized wave condenses to point-particle-like scales—as been taken over by math with unphysical notions of "instantaneous and discontinuous" transformations [*e.g.* 1 p.179],[2]. However, by three normal physical processes, we describe the facts without axiomatic interference. Firstly we ascribe the particle properties to the wave-group which we construct using special relativity. Secondly, we show that the phase velocity—that, in matter, both experiment and theory prove to be faster than light—is capable of resonant response with absorbing atoms or molecules. Thirdly, we employ constants of motion, such as hidden variables that are known to cause the Mössbauer effect in decay of radioactive nuclei. We include as hidden, other variables that are known but that are difficult to calculate, like phase, many-body forces, *etc.* The fact that the end result is condensation to atomic or molecular scales is consistent with resonance within the wave-group, as the most obvious solution. To understand why this solution has taken so long to become evident, we have to revise the respective roles of mathematics and physics. In the development of physical theory, computation is neither one nor the other. We will find that, by comparison with classical statistical mechanics, quantized states have the extra characteristic property of periodicity with harmonics.

What is math? The earliest and most necessary of the sciences. It is also, surely, the best studied and most certain, because its method depends on unquestionable axioms combined by theorems that follow precise rules: that is why 2+2=4 without alternative, depending only on the axioms that define integers. In modern times math has been analyzed by Frege [3][4], by Whitehead and Russell [5] and by Wittgenstein [6] among others. Russell discovered a new paradox: some classes are members of themselves; some are not. Consider the class of *all classes* not members of themselves. If it is *not* a member of itself, then it *is*; and if it *is* a member of itself, then it is *not*. Subsequently, Gödel proved, by mapping propositions onto the infinite series of prime numbers [7] that any axiomatic system that is consistent cannot be complete[1]: it is always possible to construct a proposition, like Russel's, that cannot be solved by existing axioms. In mathematics, the axioms are interdependent: a change in any axiom, calls into question all the others. Mathematicians "choose" their axioms [*e.g.* 8 p.82 line 2: they are not trained for physics]; physicists falsify them. The method from axiom to theorem is *a priori.* In empirical physics by contrast, it is possible for contradictory theories to be equally 'true' before or after testing. That is why no physicist is justified in defensive behavior when comparing different theories. In the logic of physics, what matters is falsification of either theory. The wave theory of the electron therefore, and the particle theory may both be 'logically true' if each is unfalsified. The wave-group is simpler: the two theories are united.

The logic of physics stands in direct opposition to math, despite depending on numbers as a tool. The author of electromagnetic theory described exact science:

"All mathematical sciences are founded on relations between physical laws and laws of numbers, so that the art of exact science is to reduce the problems of nature to the determination of quantities by operations with numbers [9]."

The meaning of a proposition is its method of verification. A proposition that is unfalsifiable is meaningless [10][11][12]. New propositions are, in principle, true until falsified by empirical measurement. By comparison, math relates to physics as did Newton's calculus to his mechanics and gravity. His calculus is mathematically described with logical certainty; but an axiomatic theory is not empirically falsifiable, and therefore is physically meaningless. A negative test for a proposition in physics negates it as certainly as a contradiction or inconsistency negates a false theorem in math.

Is quantum theory (QT) meaningless? Popper's schism, in *Quantum theory and the schism in physics* [11], implies that it is at least problematic. The schism has outlasted the Einstein [13] - Bohr [14] controversy [15] and is still, in 2022, alive in prize committees. That controversy involved entanglement and disentanglement of two like particles, but the roots of the controversy are profound and will be briefly discussed in the following pages. The roots involve stability of references frames and therefore of conservation

---

[1] The wave function therefore is in principle incomplete, contrary to the Copenhagen interpretation of quantum mechanics which does not describe internal motion, to be described. When completeness refers only to the members of a closed group solution to a Hermitian matrix equation, then that is a weaker meaning. Einstein disagreed with Bohr: "The wave function does not provide a complete description of physical reality…" [15].

laws, or whether QT calculations trump them. If so, the calculations would reconstruct reality.

Feynman said of quantum theory, It is "crazy. It does not explain anything. It just gives numbers [16]. Moreover, he said, "I think I can safely say that nobody understands quantum mechanics [17]." Of course, what cannot be explained cannot be understood. What kind of physics is QT? Feynman's view is especially authoritative because, on the one hand, his 3 volume *Lectures in Physics* is a model exposition of physical understanding; while on the other hand, his prize for calculating elementary particle phenomena celebrated outstanding achievement: while contemporaneous probabilities add, serial probabilities multiply; the former increase, with number; the latter decrease because all probabilities are fractional. "You may not like it; but that's how it is" [16].

Besides entanglement, many other anomalies are presented by quantum theory including wave-particle duality; the 'point particle'; Uncertainty Principles; collapse of the wave-packet; measurement, probability amplitudes versus conservation laws, quantization of gravity *etc.* A logic for orthodox quantum theory has been proposed that is a mixture of both axiom and fact: math plus physics [cf. 1]. Contrary to established physics, measurement in QT is supposed to be instantaneous and discontinuous—this can only be theoretical decay since what physical measurement is there anywhere that could fit this description? Moreover, it is inconsistent in QT for measurement that is faster than uncertainty in time. Since all that QT gives is numbers, it must be assumed that its 'measurements' are 'numerical and computational'? As such, we have three separate logical systems for three separate disciplines: math, physics and QT. Then a mixed theory, that shares the differentiated certainties of neither math nor physics, would bear the seed for uncertainty and schism, especially when it is uncertain which discipline is being followed at any time. The theory could be mythical.

The 21st Century is an age of enlightenment: we are all schismatics now. This is due to the internet, that includes recorded interviews and lectures that have recently become readily available. Our own description of the internal motion in the electron that is mathematically derived from special relativity; that is explanatory; and that is based on extensive, electron microscopic evidence [18] shows that the wave is not a point; that measurement is not instantaneously discontinuous [19]; and that condensation of the wave-group is a necessary and natural consequence of the wave-group energy being absorbed by an atom or molecule. However, this description does not necessarily imply that there is only one way in which experimental outcomes can be predicted; on the contrary, only that different logical systems give independent results with varying credibility. Time is Newtonian in the rest frame of a massive particle where the wavelength tends to infinity; but cannot, in physics, be normally instantaneous during measurement that is always made within uncertainties in time and space. In physics, the fact results from uncertainties in wave-groups of matter and in wave-groups of force. Mathematics is creatively and axiomatically freer. For what it might be worth, you could even construct mathematics for point particles.

There are further differences between math and physics. In empirical science, every new type of event demands a physical explanation, whether this is intuited vaguely and speculatively or rationalized in computed theory; by contrast, math enables computation in which probabilities are calculated as numerical products and where physical intuition is sometimes significant, or sometimes not. As examples, we next discuss the 'point particle', dispersion dynamics, the quantized photon, the quantized electron, and reduction of the wave-packet.

2. The photon

The quantum was discovered by Planck, and it won Einstein a prize for the photo-electric effect. On one hand the quantum explained the ultraviolet collapse in the emission spectrum of back-body radiation; while, on the other hand, the energy E of monochromatic light waves, that was imparted to photoelectrons emitted from a solid surface, was found to depend *not simply* on the intensity of an incident light beam (in free space $E=(E^2/2\varepsilon_0+B^2/2\mu_0)$)(see footnote[2]), but more directly on the frequency f of individual photons $E=hf$, where h is Planck's constant. The physical reason for the equality will be found later, in the symmetries involved in photon creation, annihilation, *etc.* [20][21]. Examples are the self-symmetries, *in time,* of Schrödinger's spherically harmonic eigenstates—due to particles orbiting in time and in phase whether they are supposed to be point particles or continuous waves—or (as will be discussed below) symmetries *in space* like crystalline diffraction in momentum space; or like dual logarithmic diffraction from quasicrystals [22].

Unlike line spectra from excited atoms, black-body radiation radiates with continuous spectra over a range of frequencies that end in ultraviolet collapse. In this case, quantization is more complex, for example in radiation from Fermi metals, described consistently elsewhere. The simpler case of Bohr's atomic model or Schrödinger's solution for the hydrogen atom, makes it easy to visualize electrons, in an ionized atomic gas, relaxing from and a quantum excited state $n=2$, $l=1$, to a ground state $n=1$, $l=0$, with emission of a photon of the precise transition state energy $hf_{2,1}$. Then energy, frequency, spin and angular momentum are all conserved properties. The decaying electron, as it emits the electromagnetic photon with quantized energy, is an oscillating, electric, dipolar aerial. The wavelength λ is also quantized because emission occurs at the speed of light c and $\lambda={}^c/f$. In principle, the photons can be counted, and individual wave-groups spread, in time and in length, with uncertainties that correspond, roughly, with Heisenberg's 'limits' [23]. The most remarkable feature of the quantized photon is that its energy, integrated over time or space, exactly matches constant:

---

[2] Where E is energy scalar, E is the electric field vector and B the magnetic field intensity vector, ε_0 the permittivity of free space, and μ_0 the magnetic permittivity.

$\phi(x{=}0,t)$

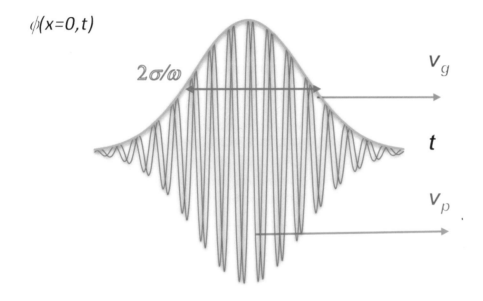

Figure 1. Normal wave packet including conservative function (orange) that envelopes infinite, responsive, elastic, complex wave (red and blue), with uncertainty $2\sigma/\overline{\omega}$ (pink double arrow) at time $x=0$. The Fourier transform of a Gaussian is Gaussian, so t may represent any of the four variables x, k_x, ω, or t. In massive particles, the envelope group velocity $v_g{<}\text{c}$ (orange, distance travelled per unit time); the phase velocity $v_p{>}c$ (blue).

$$E = \int (\boldsymbol{E}^2/2\varepsilon_0 + \boldsymbol{B}^2/2\mu_0)\mathrm{d}\tau = hf_{2,1}$$

(1)

not only because the photon wave function in space is self-harmonic over time in the initial and final electronic states, but also because it is a consequence of the wave equation [24 p.261]:

$$\frac{\partial^2 E_i}{\partial x_i{}^2} - \varepsilon_0\mu_0\frac{\partial^2 E_i}{\partial t^2} = 0$$

(2)

where the subscript i denotes the three dimensions x, y and z, and because of Maxwell's equations, including [24 p.256]:

$$\mathrm{curl}\ \boldsymbol{E} = -\frac{\partial \boldsymbol{B}}{\partial t}$$

(3)

As the rate at which an electric vector increases, the induced magnetic intensity decreases, and *vice versa.* The phase relationships of the transverse real E_y and B_z components of the electromagnetic wave can be represented by a complex wave function, using Euler's formula. This representation is extended to the complex wave function in matter.

For simplicity, represent electromagnetic waves by two of the four[3] dimensions by the normal distribution [25][26] expressed in natural units $h=c=1$ (**Figure 1**):

$$\varphi = A.\exp\left(\frac{X^2}{2\sigma^2} + X\right)$$

with imaginary: $$X = i(\overline{\omega}t - \overline{k}x)$$

(4)

where the real and imaginary parts of the exponent represent E_y and B_z respectively, *i.e.* with displacements in the transverse plane. Here, A is a normalized amplitude, σ an uncertainty, a mean angular frequency $\overline{\omega}$ and a mean wave vector $\overline{k}$, in the direction of propagation x for simplicity. Notice that displacements on the real and imaginary planes are drawn on a single plane along with the intensity of the wave. We can define uncertainty as the width $2\sigma/\overline{\omega}$ where the amplitude falls to the fractional height e^{-1} of maximum. The full wave-group would represent not only the quantized electromagnetic

---

[3] 5 dimensions when magnetic spin is included [26] but only 4 dimensions, if spin is treated separately as internal to motion.

decay from an excited hydrogen atom, but equally the radiative emission from a dipolar adio antenna. In electron photoemission, the wave is complex in the normal way.

Dirac believed that this wave-group is unstable[4] [27][28] but his notion contradicts the commonly observed microwave background whose photons have existed for 14 billion years. His opinion therefore must be a mistake. Others shared his preference for beauty over truth [29] [cf. 30]. Whatever that may be, the close correspondence between light optics [31] and electron optics [32] show that the wave-group equations (in sections 3 and 4) apply generally to both light and electron optics but with the important exception that whereas the photon is massless, the electron has the quantum mass m_e. Moreover, the displacement is in the complex, transverse plane and is not polarized as is light in the vertical, horizontal, right circular and left circular polarizations. The wave-group in **figure 1** unites the crowning achievement of 19th century physics, namely the combination of electromagnetic wave-particle duality, with the novelty of the quantum at the turn of the 20th century.

Besides demonstrating the duality without separating particles (the center of the wave-group) from their waves, the wave-group explains many of the absurdities that follow from the concept of a 'point particle', including collapse of the wave packet [2]. The absurdities mostly disappear after physical analysis [1], that is discussed in due course, along with the problematic logic of computation.

Important consequences of equation 4 occur in relativity and will be summarized in the next section. The consequences are due to the inclusion of internal motion that is neglected in Dirac's relativistic equation. That equation has widespread use, including

―――――――――――――

[4] Or perhaps his belief was a self-serving short cut. Dirac believed that truth is impossible; all he claimed was beauty [28]. His substitute is not incontrovertible in the way of Descartes' *cogito*; nor convincing as axiomatic logic in math; nor again as empirical falsification in Popper. By comparison, Dirac's 'beauty' is obviously subjective; besides being ugly as lies when it is untrue (falsified) and simplistic. You can call math beautiful if you are obsessive; a proposition in physics is either true or false. He thought Einstein's 'reality' was irrational and religious. However, those who claim that the theory of evolution replaces God [29], beg the question: evolution cannot be god if He created it (the fact of evolution counters only the argument from design (*e.g.* in Aquinas' *Summa contra gentiles*). If Einstein's god is the common intuition about a unique cause and purpose, that is outside the range of the logic of either math or physics. Pre-science begins with intuition. "The moral is ... you must not monkey with the creed [30]". Without initial intuition, how does theory start?

spin, helicity *etc.*, but it has left opportunities for the schisms mentioned earlier[5] [33]. In time and space, the wave function ϕ, in equation 4, is typically quantized. Then, whereas some physical variables, such as unmeasured phase, are typically ignored in QT; in diffraction experiments, such as Young's slits, phase is carefully conserved with its dependence on translation. Phase is also evident in figure 1. In sections that follow, we will discuss the momentum of a massless photon in wave mechanics and surprising phase coherence that is found in scattering by hierarchic materials in logarithmic space.

---

[5] For example, in relativizing the equations of quantum mechanics. Schrödinger's solutions for the hydrogen atom are non-relativistic because they ignore mass increase with relativistic motion; however, approximations can sometimes be made by calculating space contractions and time dilations between given events [33]

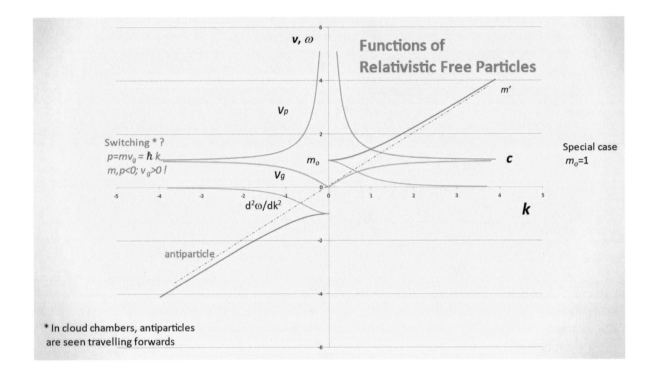

Figure 2. Functions of relativistic free particles (equation 8). To avoid unphysical singularities when $k=m_o c$ [34], the antiparticle is given negative mass, positive velocity and negative momentum. The antiparticle travels forward in time, as in cloud chambers. The dynamics of the photon, with $m_o=0$, is represented by the dash-dot line.

3. The relativistic wave-group

The speed of light is invariant in all inertial reference frames. Einstein proved an important consequence [35]:

$$E = m'c^2 = \gamma m_o c^2$$

(5)

where m' is the relativistic mass and γ the Lorentz factor. In terms of rest mass m_o:

$$E^2 = \mathbf{p}^2 c^2 + m_o^2 c^4$$

(6)

where the momentum $\mathbf{p}$ is three dimensional, in space. We can write. like de Broglie for the direction of propagation, $p = h/\lambda = \hbar k = m v_g$, by noting that his velocity must be the wave-group velocity [18]. Likewise, from Planck's law $E = \hbar \omega$, using the reduced Planck constant. In natural units, equation 6 gives a true meaning for rest mass:

$$m_o^2 c^4 / \hbar^2 = \omega^2 - k^2 c^2$$

(7)

By working in two dimensions with these substitutions for simplicity, and after differentiation of equation 6, with respect to k, it follows that the product of the phase velocity $v_p = \omega/k$ and group velocity $v_g = d\omega/dk$ is equal to c^2, since, from relativity (**figure 2**):

$$\frac{\omega}{k} \cdot \frac{d\omega}{dk} = c^2$$

(8)

It follows further that, in matter, the normalized phase velocity, v_p/c, is the inverse of normalized v_g/c. The former is, in free space, greater than the speed of light, since the group velocity $v_g < c$ is the familiar velocity described in special relativity. The massless photon has $v_p = v_g = c$ [6]. Experimental verification for equation 7 is so extensive in optical and electron microscopies that it is irrefutable [18] (**figures 3 & 4**).

---

[6] According to the equation, the phase of a photon, if it had non-zero residual mass, would travel with phase unlocked from the mean of the wave-group. One effect of mass is to unlock the step of the two velocities in the wave-group.

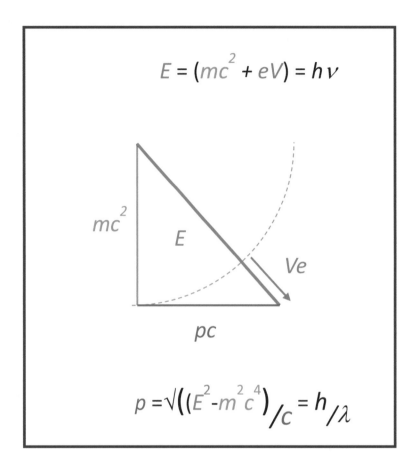

Figure 3. The frequency ν and wavelength λ of an electron microscope probe are related by Pythagoras' theorem as in relativity (eq. 7). The probe energy is the electron rest mass energy plus the accelerating energy of the gun.

Equation 7 provides interesting confirmation of wave-particle duality from the perspective of special relativity. This is more easily recognized by employing natural units $c=1=\hbar$. Then:

$$m_o = \sqrt{(\omega+k)(\omega-k)}\,.$$

(9)

The first bracket is conservative and describes the macroscopic properties: energy, momentum *etc.,* which are particle like; while the second bracket describes the wave properties that are evident in diffraction and are important in atomic structures and interactions. The two brackets therefore correlate on one hand with the group in figure 1, and on the other with internal phases whose harmonies quantize the properties of light waves and matter. Equation 9 also defines mass as a store of energy, weighed by its energetic significance but restrained in its velocity of motion. These qualities we will examine further in later sections.

Notice that the Dispersion Dynamics described by equation 5 to 9 is directly derived from many times verified Special Relativity. It is also consistent with the wave-group, the quantum, condensation and many other phenomena that are anomalous in 'orthodox' theory. A further tutorial explanation for the differential group velocity $d\omega/dk$ is given in Appendix 1.

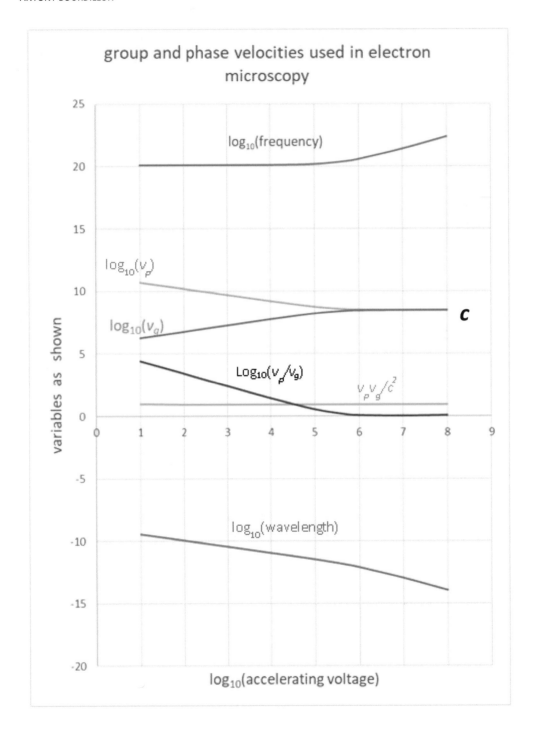

group and phase velocities used in electron microscopy

←**Figure 4.** Plots of values calculated for parameters in various electron probes against accelerating at voltage/*keV* [18], including from top down: the frequency (blue) in SI units; the phase velocity v_p (green), the group velocity v_g (purple); the ratio of phase/group velocities (navy blue); the product of phase with group velocities (yellow); and the wavelength (red). Notice the systematic relativistic changes when $Ve \simeq mc^2 \simeq 0.5$ MeV, excepting only the constant product $v_p.v_g$.

Figure 5. Simulation of 100 keV electron beam transmitted by a narrow slit. Notice the variations from near field to far field imaging in Fresnel diffraction: Δx is minimum at the critical condition, with Δp_x passing from negative (converging) to positive (diverging). In the near field, the dual uncertainty is 4 x larger than at Heisenberg's 'limit' and 20x larger in far field. Simulation due to C.B.Boothroyd [36][37]. This image illustrates how the coherent, monochromatic, incident beam interacts with the absorbent slit.

4. Uncertainty

Heisenberg's Uncertainty Principles are often treated as so basic in quantum mechanics that they are called axiomatic. However, since physics is not axiomatic, the principles belong to mathematical quantum theory and are not tested for 'truth'. Problems occur that are both theoretical and empirical. The first is the meaning and application of his 'limit'. Secondly, the uncertainty can sometimes be accurately calculated from the wave-group or experimental arrangement according to established techniques in wave particle duality [37]. Thirdly, the Heisenberg limit is frequently an order of magnitude smaller than we calculate in physical arguments. We proceed therefore with a physical derivation to compare with Heisenberg's limit and follow it by references to experimental work (**figure 5**).

Begin by writing from equation 4, the intensity:

$$\phi(t)^*{}_{x=0}.\phi(t)_{x=0} = A^*A.$$

(10)

Then, ignoring the normalized constants, the uncertainty in position when $x=0$ and the argument in equation 4 is $\pm$unity:

$$\Delta t = \frac{2\sigma}{\omega}$$

(11)

The Fourier transform of a Gaussian is Gaussian. The transform of equation 4 is:

$$F(\phi(t)^*{}_{x=0}.\phi(t)_{x=0}) = \sqrt{\frac{\sigma}{\omega\sqrt{2}}} \exp\left(-\frac{\sigma^2\omega^2}{4\,\overline{\omega}^2}\right)$$

(12)

and, as before, when the argument of the exponent is ± 1 the uncertainty in angular frequency is given by:

$$\Delta\omega = 4\overline{\omega}/\sigma$$

(13)

The dual uncertainty, after σ and $\overline{\omega}$ cancel, yields:

$$\Delta t.\Delta\omega = 8$$

(14)

Notice that the simplicity of this formula is consistent with the wave-particle origin of the important physical property. This *expected* value is 16 times greater than the *'limit'* under Heisenberg's axiom:

$$\Delta t . \Delta E \geq \hbar/2$$

(15)

In space, the argument corresponding to equation 14 yields:

$$\Delta x_i . \Delta k_i = 8 , \quad i = x, y, z$$

(16)

In our experiments in micro-analysis [38][39] and in nanolithography [38][37], we have found that semi-classical calculations are consistent with equation 14 rather than with the axiom in equation 15. Moreover, wave mechanical effects sometimes reduce the uncertainty in near field imaging (**figure 5**). These effects are represented in neither equation 14 nor 15. Furthermore, there is a large margin of error given by the inequality sign in equation 15, so the equation is not strictly wrong, but its information content is comparatively small. In mathematics, that hardly matters, but some branches of physics may, in consequence, be questioned.

Returning to figure 5, where the Critical Condition in X-ray lithography occurs below the cross-over with zero intensity on axis, it appears that the dual uncertainty crosses from negative to positive. This data is well verified [31] and is in obvious conflict with Heisenberg's 'limit'.

5. The point of a particle and its momentum

How many angels can be placed on the point of a needle? As many point particles as in quantum theory before Pauli exclusion. From Schrödinger's theory, Bohr's radius a_o of the hydrogen atom is easily derived and found to be consistent with measurement of atomic sizes *etc.* Further input is needed to derive magnetic radius $a_o\alpha$ [26] where α is the fine structure constant, and to derive its electrostatic radius a_o/α^2 [40]. By assuming that the point particle is axiomatic we can replace it by the center of mass of the wave-group. This is done to represent the close similarities between light optics and electron optics. When we measure the location of an electron, *we* are measuring the center of its wave-group, and this is the center of its mass.

The 21st Century is an age of enlightenment: we are all schismatics now, owing to the freedom of the internet. Our description of the internal motion in the electron that is derived from special relativity, and that is based on extensive, electron microscopic evidence [18], shows that the wave is not a point, and measurement is not instantaneously discontinuous [1]. Consider momentum: classically, the momentum p of a particle is its quantity of velocity, $p=m_o v$. However, the photon has zero rest mass. Consider a plane wave on a water surface with a small bubble that is located at some point by surface tension (**Figure 6**). Notice that, because of cyclic motion within, the wave is energetically economic compared with what the movement of a flat surface slab would be. This is a well-known concept in material science where a ripple in a carpet translates more easily than does the whole carpet and a dislocation travels comparatively easily in a slab of perfect crystal. Moreover, in the wave-mechanical interpretation of equation 4, simplified for the simple case $t=0$ in natural units:

$$\varphi = A.\exp\left(\frac{-(kx)^2}{2\sigma^2}\right)\exp(ikx)$$

$$(17)$$

where

$$<\phi^*.\phi> = 1$$

$$(18)$$

Figure 6. A plane wave with wavelength λ, oscillating with frequency f, travels with phase velocity $v_p = \pm\ \lambda.f$. The bubble (filled circle) executes oscillatory motion with zero net transport. Supposing every point on the surface is also oscillatory, what is the momentum of the wave?

So that, by applying the chain rule, the partial derivative:

$$\frac{\partial \phi}{\partial x} = A.\left(\frac{-k^2 x}{\sigma^2} + ik\right) \exp\left(\frac{(-kx)^2}{2\sigma^2} + ikx\right)$$

(19)

At time $t=0$, the formula simplifies to:

$$\hbar < \phi^*.\frac{\partial \phi}{\partial x}> = -i\hbar k \ .$$

(20)

Since x is antisymmetric about the origin while $\phi*\phi=1$ is symmetric, the first bracketed fraction in equation 19 evaluates to zero. Heisenberg's momentum operator $\acute{p} = \hbar/_i \cdot \partial \phi/\partial x$ produces a similar result [7] (confirming the old German adage, "We believe in Heisenberg, and calculate with Schrödinger.") The point is that, in wave mechanics, the motion is shifted by 90 degrees in complex space when the bubble trajectory oscillates out of phase with the wave function $\phi \sim \exp(ikx)$ in (**figure 6**). The motion is maximum when the real wavefunction displacement passes zero. The momentum is a constant wave-group value (equation 20), and so depends on v_g; not the phase velocity v_p. In correspondence, de Broglie's momentum depends, not on λ; but on its inverse λ^{-1}, consistent with equations 8 and 20, when the photon has finite momentum $i\hbar k$, even though its rest mass is zero:

$$|p| = m'v_g = \frac{m'c^2}{v_p} = m'c^2.\frac{\hbar k}{\hbar \omega} = \frac{|h|}{|\lambda|}$$

(21)

The greater the momentum, the shorter the wavelength and the higher the frequency. This relation is not hypothetical; but is derived from the wave-group and wave mechanics (appendix 3).

These relationships are more obvious in the free particle wave than when they are learned from bound atomic states, where the relationships are assumed. Rest mass, for example, is not necessary in momentum; the mass differentiates the phase velocity of a particle from its group velocity. In wave theory, momentum is not in-phase resistance,

---

[7] confirming the old German adage, "We believe in Heisenberg, and calculate with Schrödinger

but out-of-phase impedance to motion, like inductance in electric circuits.

Further importance for our free particle representation becomes evident as physical harmonies in waves are found necessary for quantum effects generally, and particularly in atomic structure and quasicrystal diffraction, as we shall see. In the latter case, wave-harmonic constraints are both linear and geometric, and both integral and irrational. This is how the periodic wave interacts with the geometric, hierarchic structure.

6. Quanta

Conciseness in mathematical quantum theory is gained by objectifying core notions. When these lose their physical origin, they lose also their causative significance and their opportunity for further development. We have seen that uncertainty is one of these notions, the point-particle, is another, so also internal motion in the wave-group. In this section we describe the physics that cause quantization in the wave-group. The method is to notice firstly that many properties are quantized besides energy. In the simple case of *crystal diffraction*, the underlying cause of the momentum quantization is evident: it is the harmonies that occur between periodic structures, periodic probes, and periodic scattering patterns. Typically in crystals, any Bragg scattering event from one atom is in phase with the scattering from the other atoms. Surprisingly as we shall see, the same condition occurs in *quasicrystals* where the probe and scatterer and diffraction space are not all periodic as they were in crystal diffraction. But the harmonies that occur in dual space are complicated. They are necessary to produce the quantum effects that are observed. It is obvious that these effects are related to the *spherical harmonics in atomic structures* when the quanta are energetic in transitions between initial and final eigenstates. Then wave-harmonics are time dependent with quantal frequencies regulating stable orbital periods. There are, however, important differences between atomic transitions in time and *crystal diffraction* in space. Here, the individual atomic scatterings are separated by wavelength periods that ensure they occur—typically—at unique phase. In crystal *diffraction*, the quanta are momentum quanta because the scattering of monochromatic beams is approximately elastic.

The quasicrystal has remarkable, new properties that demonstrate the importance of harmony. The properties were so unexpected and peculiar, that they directly demonstrate the origin of the quantum as being the harmonies that occur in the probe's wave-group. The quasicrystal structure is hierarchic, the probe is conventionally periodic, and the diffraction scattering occurs in geometric space. It is remarkable that the diffraction occurs in strict order; but it is most remarkable how extreme this order is. At first intuition, the scattering should be diffuse as from amorphous materials. It was, for a long time, unexplained. Now we know there is a log-lin metric that is well understood, and that enables the harmonies to occur in both logarithmic and linear spaces. This fact fixes harmony as the physical basis for quantization generally.

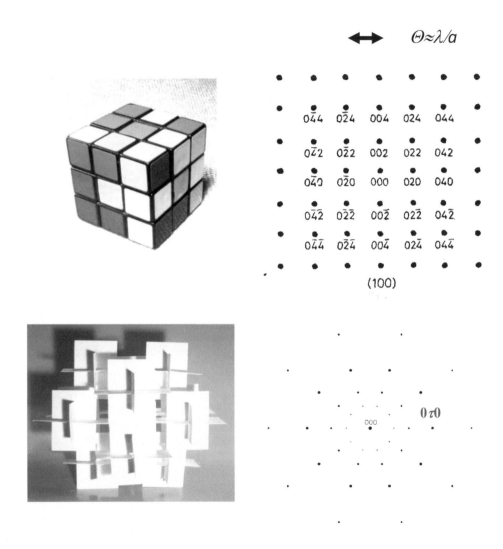

$\Theta \approx \lambda/a$

(100)

$0\tau0$

000

Figure 7. *upper figures:* cubic structure and face-centered cubic diffraction pattern normal to the [100] axis. The probe, the crystal and the pattern are all periodic. *Lower figures:* the hierarchic quasicrystal and corresponding diffraction pattern normal to the fivefold [τ01] axis (see below). The pattern is icosahedral in geometric space. The pattern and structural cells are in geometric series, with stretching factor between orders τ^2.

Figure 7 illustrates the chief difference between crystal diffraction in the upper row and quasicrystal diffraction in the lower row. In crystal diffraction, Bragg's law requires that the scattering angle, $\Theta \approx 2\theta$, corresponds to:

$$\sin(\theta) = \frac{n\lambda}{2d}$$

(22)

where the order n *is integral*, θ is the Bragg angle, and d is the interplanar spacing that are further discussed below. (In the particular instance shown, where the angles are small in electron microscopy, d_{100} is equal to half the lattice parameter a). In crystals, the atoms are periodically ordered; in quasicrystals, clusters of atoms are arranged in hierachic arrays, that are therefore geometrically oredered.

Since the ordering is wrong, we know that quasicrystals do not obey Bragg's law; the order must be expressed in a different way and the variables must be redefined. The quasi-Bragg angle θ', that will be defined after quasi-structure factors are calculated, is supposed to follow:

$$\sin(\theta') = \frac{\tau^m\lambda}{2d}$$ '

(23)

where $\tau = (1+\sqrt{5})/2$ is the golden section, and primes indicate quasicrystal values, initially undefined. The order m is integral, like n, but it defines an irrational series.

The cells shown in figure 7 (lower row) are each icosahedral and can represent any member of the hierarchic series. The unit cell is likewise icosahedral. It contains a comparatively small *Mn* atom surrounded by 12 larger *Al* atoms with atomic diameters that yield an extremely dense cell structure. Since the *Al* atoms are generally edge-sharing, the stoichiometry is 1:6. This was the composition of the earliest quasicrystal discovered [41]. Each corner of the golden triad representing the unit cell is the site of an *Al* atom, typically edge-sharing. In higher orders, the corners are sites of subclusters. The dense unit cell is the apparent driving force for the structure. The stretching factor between orders is τ^2.

Such sites are imaged by Bursill and Peng [44] in **figure 8** where a hierarchy of structures can be seen. The image is in reverse contrast, so that cell centers, including *Mn* atoms, are in dark contrast. The red circles in the image mark, with increasing diameter, a *Mn* atom, a unit cell, a cluster and a supercluster order 2. This contains about 1000 atoms,

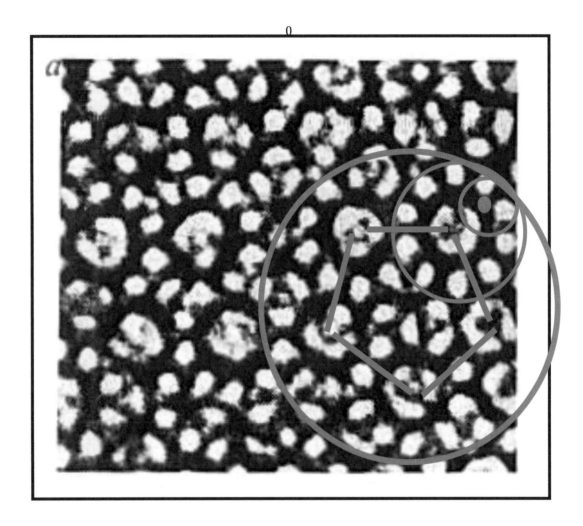

Figure 8. Transmission electron microscope image of icosahedral Al_6Mn observed in phase contrast, optimum defocus. The incident beam is parallel to the five-fold [τ01] axis of the thin foil. The reverse-contrast image includes a supercluster order 2. The raw image is due to Bursill and Peng [44] re-published with permission.

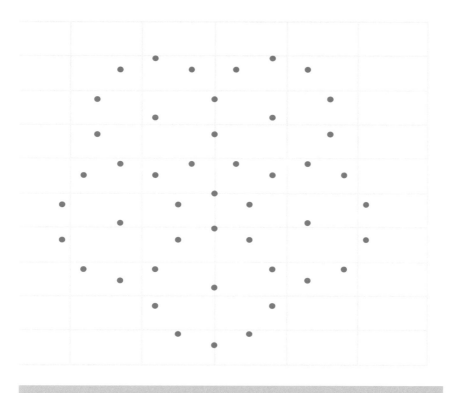

Simulated
Cluster peripheries in**reverse** contrast

Figure 9. Simulations of low-density regions near to central plane of a supercluster, as imaged in figure 8. There, unit cell centers that site heavy *Mn* atoms lie in dark regions. Off-plane, corresponding patterns become less symmetric, as they do in the micrograph [42] beyond the supercluster. There is little control over the exact plane that is polished and imaged.

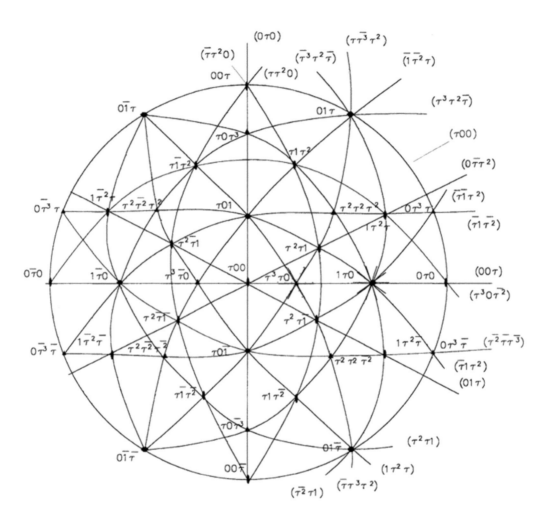

Figure 10. Stereogram of principal axes of an icosahedral structure, indexed geometrically. Normal to the principal axes are the principal diffraction planes, following corresponding conventions in crystallography [43]. The central axis is [τ00]; the outer plane is (τ00).

but there are fewer in the thin section. The difficulty of electropolishing specimens of defined planes is the reason why a wider supercluster image is not available. However off-plane simulations corroborate the interpretation [42].

planes is the reason why a wider supercluster image is not available. However off-plane simulations corroborate the interpretation [42].

The simulation in **figure 9** marks, with blue spots, the low density regions in the reverse contrast image. The model structure matches the scale of the microscope image.

Knowing the structure we need to simulate the diffraction pattern. To do so, it is necessary to understand the Quasi-Bragg law and, if possible, to verify it. The first step is to index the pattern by taking hints from the methods of crystallography about how this should be done. We noticed at the outset, that the dominating fact in the diffraction pattern is its geometric order with factor τ^{2m}. The stereogram in **figure 10** graphs the prical axes, and diffraction planes normal to them. This stereogram is used to index the patterns [43], based only on the identification of icosahedral structure. The principal axial patterns were completely indexed in 3 dimensions and reflection intensities calculated. We borrow Ockham's razor to notice that dimensions should not be multiplied without necessity: not 6 dimensions; not 8 dimensions. Knowing both the structure and correct indexation, the diffraction pattern was simulated and this revealed the reason why the non-periodic structure diffracts with a sharpness typical of crystals; not at all like amorphous glasses nor disordered solids.We cannot use Bragg's law because it has the wrong order;but we are equipped to apply the *structure factor* method. This is the method used to simulate, in crystallography, the intensity variations that are due to scattering atom sites in their unit cell. It is also used to identify diffraction beams that are forbidden when, for example, half of the atoms in a unit cell scatter in antiphase with the other half. The method is sensitive to other structural features including disorder.

Before moving to simulation, it is necessary to the understand more detail about how the diffraction occurs. In crystals, diffracting at a Bragg condition *hkl,* the interplanar spacing d_{hkl} is unique for all scattered beams at a particular condition. It is easy to see how the scattering of the periodic probe can occur *in-phase* at every atom periodicallty arranged in 3 dimensions (**Figure 11**, left). After scattering, the incident beam emerges as an in-phase plane wave front. The figure illustrates two diffraction orders: a first order in red, and a second in green with a larger scattering angle. In first order, the path lengths of rays reflecting off adjacent planes differ by λ; in second order, they differ by 2λ, *etc.*

The quasicrystal is more complicated: there are multiple inerplanar spacings (figure 8 and figure 11, right): the emerging wavefront is bent and twisted. It is even surprising that

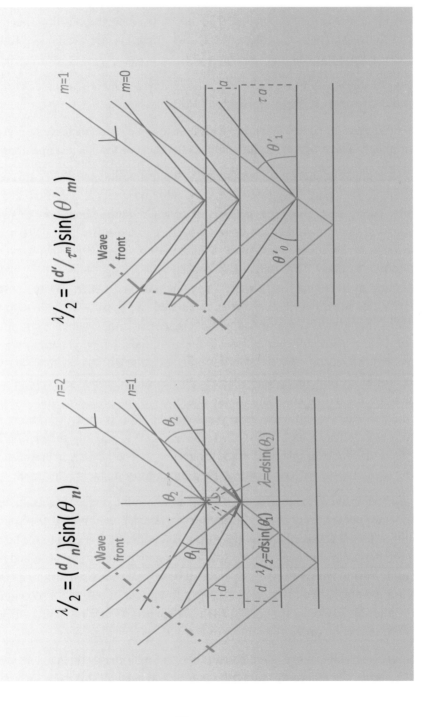

Harmonic Bragg law v Hierarchic Diffraction

← **Figure 11.** Comparison of crystal scattering having unique interatomic spacing, *d.* In Bragg diffraction, path lengths differ in all coherent rays with path differences an integral number of wavelengths; in quasicrystal scattering having multiple interatomic spacings, (including *a* and *τa*), it is surprising that diffracted beams cohere with *effective* spacings $d'=c_sd$. In this figure the 'atomic planes' are imaginary cin quasicrystals; the real answers are computational in 3-dimensions. The diffracted wavefront is then, in complex summation, linear.

diffraction occurs at all in an ordered way. Remarkably, scattering factor calculations mimic the sharp diffraction pattern. By accounting for special features in the quasicrystal structure, they reveal new light on hierarchic icosahedra.

In a crystal, the structure factor of an indexed diffraction beam h_{hkl}, *i.e.* in 3 dimensions, is calculated by adding the scattering power due to all the atoms in the *unit cell*. The relative scattering of these atoms depends on individual atomic scattering powers, and to the locations of the atoms relative to the incident scattering beam. The last of these governs the phase of a scattered beam and the intensity due to the amplitudes of all the scattered rays added together.

For a *crystal* therefore, write for the structure factors, indexed *hkl* in the mormal way:

$$S_{hkl} = \Sigma_i^{\text{unit cell}} f_i.\cos(2\pi\,(\boldsymbol{h}_{hkl}\cdot\boldsymbol{r}_i))$$

(24)

Where $\boldsymbol{r}_i$ is the location of atom *i* having scattering factor f_i, so that $\boldsymbol{h.r}$ is the projection of the atomic site onto the plane normal corresponding to the scattering plane of the indexed beam *hkl*.

For the *quasicrystal*, we adjust the structure factor to take account of two special features. Firstly, because we do not have a unit cell that repeats periodically, we have to sum over the whole quasicrystal. For practical reasons, including computer memory, byte size, and rounding errors, **Figure 12** was calculated for a supercluster order 6. Secondly, Because there are multiple interplanar spacings, we include a coherence factor. Initially it was unknown, but after it was calculated statistically to high precision, it was found to have the theoretical value of $\tau^{-1/2} = 1/c_s$ (see below), and most remarkably of all, is identical for all the beams reflected in the diffraction pattern. The value of $1/c_s$ is 11.18034…%, *i.e.* the ratio factor that increases the scattering angle compared to the Bragg angle for a known *d*-spacing and wavelength λ.

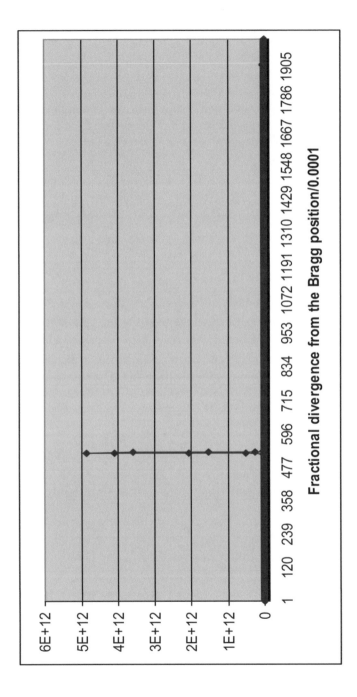

Figure 12. Quasi-structure factor for the (τ00) diffracted beam due to a supercluster order 6, obtained by scanning coherence factor c_s. Notice firstly, there is no Bragg diffraction at the origin where c_s=1; and secondly the very sharp diffraction that occurs at the quasi Bragg condition with a width that is statistical, and with the divergence (from Bragg) that is the same as the theoretical value (the irrational part of the index).

These two adjustments lead to the quasi-structure factor:

$$S^p{}_{hkl} = S^{p-1}{}_{hkl} \, \Sigma_{cc}{}^{12} \, \cos(2\pi c_s(\tau^{2p}\boldsymbol{h}_{hkl}\cdot\boldsymbol{r}_{cc}))$$

(25)

where the changes in the formula from the crystal structure factor (eq.24) are differentiated in red. The coherence factor is c_s. The formula is iterative from order p-1 to order p, with the stretching factor τ^2 that scales between the unit cell and cluster, and between a cluster and supercluster and so on. The supercluster (cluster of clusters) order is p=1,2,3... Notice that the coordinates for subcluster centers r_{cc}, substitute for r_i in equation 23. For the unit cell, equation 23 is used for the quasicrystal with inclusion of c_s beside 2π. All solids have defects, in our case, they are accounted probabilistically.

The chief and typical result of the calculation is shown in **figure 12** for the square of the quasi-structure factor of the (τoo) line in the supercluster order 6. The width of the line is not due to fuzzy phase, but is statistical. This is the most remarkable feature of the structure. The computed structure factors are the only way that we have to understand the extremely coherent spectra that are found in the diffraction patterns.

With realiosm: The sharp coherence of the quasi-structure factor in the hierchic icosahedron is a new and significant result for modern physics. It represents a paradigm shift, as in [45], for a structure that intuitively would have suggested diffuse diffraction from a disordered solid.

The paradigm shift is that it confirms the nature of the quantum, through its extraordinary harmonic mechanism for diffraction that will be described as follows. Meanwhile, 'intuition' has a pre-falsification and post-theoretical role in the evolution of physical understanding. This role is similar in moral thinking [46]. For the calculation at hand, the measured value of the quasi-lattice parameter was used. It is obtained from measured angles in the diffraction pattern, after applying the correct indexation [8] and consistent structure (**figures 13 and 14**).

The method of calculation is consistently confirmed not only by the simulation of c_s with realistic quasi-lattice parameter, but also to the measurement of diffraction line intensites that also match experimental diffraction patterns [8]. The quasi-structure factors have a different

---

[8] Consider the diffraction pattern in figure 7, lower row. The correct indexation is given in order from the zero order line along the horizontal: 000, 0$\tau^1$0, 010, 0τ0 (third bright ring in figure 7), 0$\tau^2$0. Measured on the (0τ0) line, the quasi-lattice parameter a=.296$\pm$.05 nm. This is larger (by a small fraction) than the diameter of Al that equals the width of the unit cell (**figure 13**, lower).

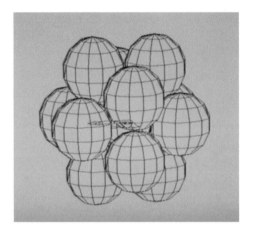

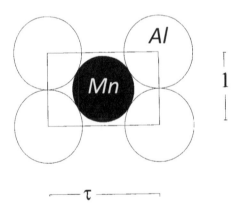

Figure 13. (*upper*) hard sphere model of the unit cell with (*lower*) cross-section a central golden rectangle. The cell is densely packed with quasi-lattice parameter equal to 1 on this scale, and measured to be 0.29 nm, *i.e.* the diameter of *Al* .

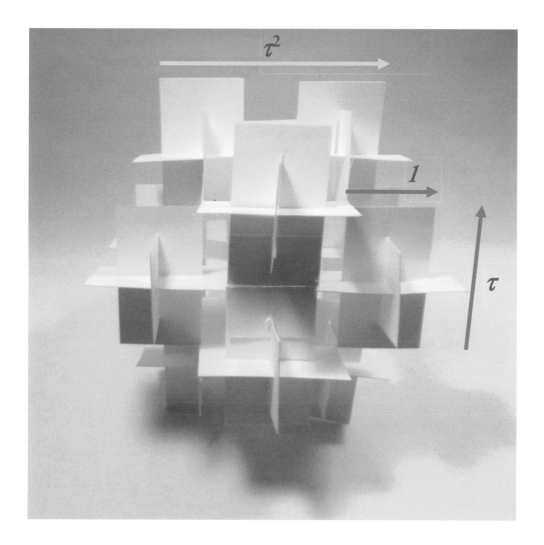

Figure 14. A hierarchic cluster of hierarchic unit cells with golden rectangle dimensions 1 x τ and stretching factor $1:\tau^2$

role in quasicrystal diffraction than do structure factors in crstalline metals. In pure face centered cubice *Al*, forbidden lines (*e.g* figure 7, upper pattern) frequently occur when two atoms in the unit cell scatter in antiphase with two others: Then $S_{hkl}=0$; whereas in the quasicrystal, the integral orders (n>1) are mostly inhibited. For the quasicrystal, most of the structure factors from non-equivalent indexed diffraction beams are differentiated and significant. The values of the various structure factors vary widely.

So far an interpretation of the geometric series diffraction pattern has been described. The method is consistent, internally with diffraction and imaging data and also with atomic dimensions. We now have to understand the coherence factor that was used as a scanned parameter, in figure 12, to discover the diffraction condition in the hierarchic, icosahedral structure. In particular, we have to understand how a periodic probe interacts with the irrationally ordered structure to diffract into geometric space. We start by noticing that the following Fibonacci sequence [47] is both linear and geometric with both rational and irrational numbers:

$$\tau^m = F_m(1,\tau) = \partial_{(m,1)} + F_{m+1}(0,1) + F_m(0,1)\tau$$

(26)

as is easily proved for integer *m* by mathematical induction. The first term on the right hand side is the Dirac delta function. Of the second and third terms, one is rational, the other irrational. The last can be approximately rationalized by substituting for τ, the rational fraction $^3/_2$. Then an irrational residue IR can be extracted by subtracting the rational approximation from the original geometric term:

$$IR = \tau^m - (F_{m+1}(0,1) + F_m(0,1)3/2)$$

(27)

and we can define a metric function:

$$m_f = 1 + \tau^m\text{-}IR = 1 + \frac{2\tau^m - F_{m+4}}{2 F_{m+1}}$$

(28)

$$= \frac{1}{c_s} = \tau - ^1/_2 = 1.118034\ldots \text{ for all } m$$

(29)

where the theoretical metric function equals the computed inverse of the coherence factor to an accuracy of several figures. It was demonstrated that the approximate rational indices corresponded to an imaginary solid that would diffract by Bragg's law (with $c_s=1$), and this implies that the cause of the quasicrystalline divergence from the law is due simply to the irrational part of the indices shifting the phase of the probe. It is remarkable that the shift is the same in all diffracted beams, and this implies that the divergence and coherence factor is a special property of hierachic icosahedral scatterers. With the understanding we now have, we can add techniques that are commonly used in understanding high resolution electron microscope images. In the light of the log-lin metric function, they will lead to a demonstrable understanding of what quanta truly are in physics.

How do Bloch waves behave in dual log-lin space? Before proceeding to this step, we should verify the theory just described. We do this by measuring the quasi-lattice parameter for the icosahedral structure shown in figure 7, lower row. **Table 1** summarizes the method used to measure the quasi-lattice parameter, i.e. the width of the unit cell, by applying the coherence factor to the measured scattering angle for a line of known index $(0\,\tau 0)$. The line, which belongs to the third bright ring on the diffraction pattern (see figure 7), was recorded about the $[\tau 01]$ axis. The Bragg equivalent interplanar d-spacing has the measured value 0.205 nm [48]. The measurement was calibrated on a second phase Al microcrystal. The quasi-lattice parameter has the vale $0.205\ c_s, \tau$. The value is close to the known diameter of Al (figure 13 lower), while making allowance for the peculiar electronic band structue [49] of the peculiar icosahedral intermetallic. The consistency of the method verifies the theory.

We can therefore proceed to understand details of the momentum band structure employed in the diffraction. The basic models are those regularly employed in crystallography [32], but they must be modified to take account of the geometric space containing both the hierarchic structure and the diffraction pattern [50].

The normal method involves analysis of the two beam condition, i.e. at the symmetric Bragg condition. The incident beam and reflected beam interfere. By quantum mechanics, the interference separates into two commensurate momentum states, or Bloch waves. One cosine wave state is strongly interacting and is intense at atomic scattering sites; the other sine state is denser at interatomic regions The momentum band splitting phenomena cause thickness fringes in wedge foils. In crystals, the waves oscillate from one state to the other as the electron beam passes through the specimen foil by the pendelösung effect. Thickness fringes are much studied in electron microscopy and occur also in quasicrystals [48]. The bands emerge from the lower side of a wedge foil. In high resolution imaging, Bloch waves reveal commensurate lattice planes, along with defect structures of many kinds.

Figure 15 illustrates such Bloch waves, specifically the *blue* waves in crystals. The two density waves shown translated one above the other are invariant under lateral translation on linear crystalline lattices (but not not on the geometric quasicrystalline quasi-lattice). Both blue waves are commensurate with the Bragg (200) planes as they would be in separated crystals. They arrange in periodic, linear order.

Bragg (crystals)	Hierarchic (QC)	(Comment)
$n = 2d\sin(\theta)/\lambda$	$\tau^m = 2d'\sin(\theta')/\lambda$	Harmonic laws: Bragg / Heierarchic
$S_{hkl} = \Sigma f_i \cos(2\pi \mathbf{h}_{hkl} \cdot \mathbf{r}_i)$	$S^p_{hkl} = S^{p-1}_{hkl} \sum_{cc}^{12} \cos(2\pi c_s(\tau^{2p}\mathbf{h}_{hkl}\cdot\mathbf{r}_{cc}))$ iterative	Structure factors Give c_s, a and d'
$d = a/h$	$d'_h = a\, c_s/h$ *	$c_s = \theta^{Bragg}/\theta^{qc} = \theta/\theta'$ **
	for $(0\tau0)$, $a' = 0.205\ \tau c_s$ nm $= 0.29$ nm $a^* = 2\pi c_s/a$	Measured q-lattice parameter $a' \approx$ Diameter of Al

Table 1. .Summary of measurement of quasi-lattice parameter using coherence factor with quasi-Bragg angle of $h = 0\tau0$ diffraction line: measured, analyzed, verified and complete

Figure 15 illustrates such Bloch waves, specifically the *blue* waves in crystals. The two density waves shown translated one above the other are invariant under lateral translation on linear crystalline lattices (but not not on the geometric quasicrystalline quasi-lattice). Both blue waves are commensurate with the Bragg (200) planes as they would be in separated crystals. They arrange in periodic, linear order.

When the Bloch waves are stretched by the metric function (equation 28) like the corresponding diffraction pattern in a quasicrystal (figure 12), the *red* quasi-Bloch wave is commensurate with the geometric, hierarchic structure. The two red waves in the figure have mirror symmetries about all translations $a'\tau^m$, m= -∞,-1,0,1,2,3,4… This gives a primary condition for coherent scattering by the quasicrystalline, geometrc, quasi-lattice. But notice that local linear symmetries also occur in the (red) quasi-Bloch waves about geometric intercepts. This property facilitates sharp quasi-Bragg scattering that occurs—not directly between atomic sites as in Bragg scattering from crystals—but from cluster centers of all orders, *i.e.* geometrically spaced. Thus the periodic X-ray or electron probe scatters from the geometric quasi-lattice into the geometric diffraction pattern in geometric space (figures 7 and 11).

The dual diffraction condition is not arbitrary but occurs as the consequence of the extraordinary structure factors (figure 12) that are calculated in every beam in the original diffraction pattern.

Notice that the number of cycles that occur between intercepts on the geometric quasi-lattice vary as (1),1,2,3,5,8… in Fibonacci sequence. This sequence is the denominator in equation 28. (The first of these cycles corresponds to the angular gap between the strong $(0\,\tau 0)$ line in figure 7, as indexed, and the adjacent $(\tau^0 00)$ line. The latter line corresponds to the unit cell edge width in the lower row of figure 13.

These Bloch waves may not yet have been observed directly in the exact manner of the simulation, *i.e.* with the highest resolution in quasicrystals. However, the effects of the waves in similar applications are certainly observed. In convergent beam electron diffraction for example, fine line fringes that are due to interference between quasi-Bloch waves, show that the diffraction is properly represented by appropriate adaptation of the conventional description commonly used to describe defects in crystal structures [48][50]. Other features, that are common to crystallography, are displayed in quasicrystal diffraction, such as Kikuchi lines [48] that are due to inelastically scattered electrons reflected from atomic planes. These properties are consistent with the hierarchic structure.

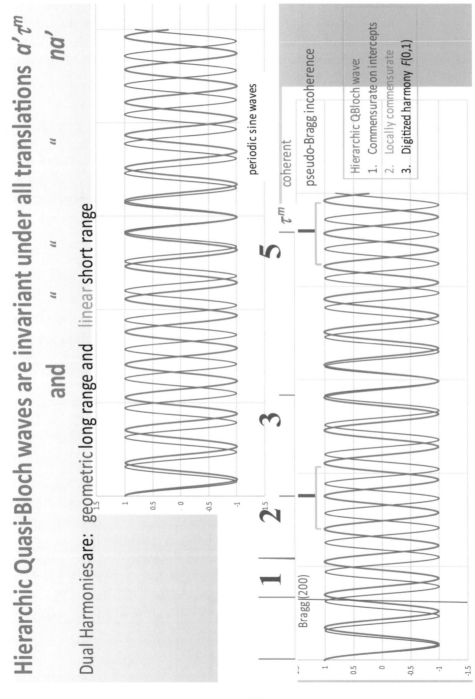

← **Figure 15.** See text: Two (blue) Bloch waves not symmetric about geometric intercepts compared with two (red) quasi-Bloch waves that are dual harmonic, both long range geometric and short rang linear about each intercept.

The foregoing description of dual harmony in quasicrystal diffraction has consequence for understanding quanta in quantum mechanics. The momentum quanta in these diffraction patterns are dual harmonics, contemporaneously, in periodic and geometric spaces. The fact leads to the inference that all *quanta are harmonic states in time and space. They are created by harmonic resonances and annihilated by harmonic interference.*

With this conclusion we are able to revisit condensation: the reduction, of a dispersed wave-function (figure 5) to atomic or molecular dimensions at detection, as in the following chapter. Meanwhile, Shechtman's definition for the quasicrystal: "a metallic phase with long range order but no translational symmetry [45] [9]" now makes sense in neither intuition nor fact. The consistent theory for both the diffraction (including indexation) and imaging has been given only by implementing the hierarchic icosahedral structure.

---

[9] Sometimes called a Dirac quasicrystal, though he died about the time of the first publication, when the structure and diffraction mechanism were were yet unknown.

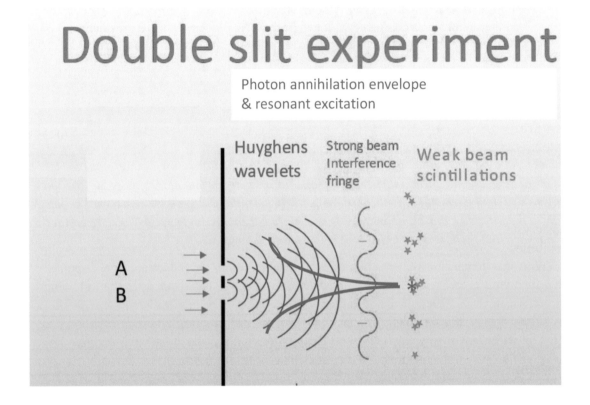

Figure 16. Diagram showing interference due to two slits, A and B, illuminated by coherent beams. The consequent diffraction patterns that are formed are either strong (red) from a laser, or weak (green) by passing throough an absorber, with photon counting. Condensation (purple) occurs in weak beam by resonance between the photon wave-group and a competition won by the hidden variables (phase coherence, emission recoil *etc.*) of the absorbing molecule (purple). By this resonance, the photon beam is condensed.

7. Condensation

The fact that collapse [2] reduces the wave front to an individual atom or individual molecule; but not to some more general spatial dimension, indicates that the target is instrumental in absorbing the energy (or momentum) of the collapse. Mysterious tricks in math are not explanations for obvious events in physics, nor are they explanations in 'orthodox' QT. What do we know?

We know that the wave-group has uncertainty ($\sigma > 0$), and that light energy is absorbed in finite time $\Delta t \simeq 2\sigma/\overline{\omega}$. In that time, the absorbing molecule annihilates the dispersed photon by anti-phase resonance with the incident photon. We know, from quasicrystal diffraction, that quanta are restrictions in motion that are due to necessary harmonies in their respective photonic and material wave structures. It is obvious that anti-phase resonance concentrates the photon dispersion in space onto the excited molecule in the absorbing scintillator (or other detector), and the excitation subsequently decays by scintillation (**Figure 16**).

 Thus, condensation is a consequence of the properties of the wave-group that, on transmission through the double slit, disperses in time and space and, during detection, concentrates its energy, momentum and other quantum phenomena onto the scintillating molecule. The concentration occurs because of necessary harmonies of structure in the dual wave-particles of the photon and absorbing molecule. The anti-phase resonance annihilates the photon.

Given what we know, this process is obvious in physical quantum mechanics. We should ask why a photon is selectively absorbed by one molecule rejecting all the others. It should be assumed that more than one molecule may begin to resonate but fail to completely absorb the photon ... and reverse their resonances by in-phase secondary decay that returns the secondary molecule to its original state. The single selection of a molecule is surely due to hidden variables. These include phase coherence between ground states in the absorbing molecule and the exciting photon, atomic emission recoil that adjusts the Poynting vector within the photon. and other experimental factors that are not measured.

This obvious understanding of an aged anomaly leads to the following reflection about whether anything went wrong; whether there was schism at all: despite their interdependence on numbers, math has different criteria and objectives from physics. The examples here illustrate a general problem in the practice of physics: it is easy in math, to postulate a point particle for computational purpose, but easier in physics to include both wave and particle properties in the single wave-group, as in the electromagnetic photon. Then it is easier still to include the same duality, that we know in the photon, to matter generally.

In math, events that cannot presently be calculated in physics, may often be represented by probabilities of unmeasured properties that may be speculated in an intuitive, and imperfectly physical, understanding. Proper physical understanding avoids weird and unphysical axioms that are instantaneous and discontinuous *etc.* The notion that a probability is a reality fits the mathematician's purpose that reality is what can be calculated; but it contradicts the physical requirement that reality is a proposition that tests 'true'. Confusion is often due to the importance of numbers in physical theory. Sometimes mathematical and physical predictions coincide, but in other instances mathematical outcomes are weird because they conflict with common experience[10] [38][16]. In case of coincidence, such as Planck's law, it does not matter which 'reality' is used to "relate the laws of numbers." However, mathematicians who mix axioms with verified physical propositions in a single system cannot have understood the limits of *their* separate logical methods; while it is reasonable for physicists to represent unmeasured variables by probability distributions, typically normal distributions, it makes little sense to call such variables 'objective'[11] probabilities [cf. 1] as if they are real entities. The positions and momenta of individual atoms in an ideal gas are often represented by statistical probabilities, but the probabilities are not multiplied entities that must be razed by Ockham.

Under the razor, a calculated amplitude cannot be an object. We understand that a probability amplitude becomes objective when its distribution is measured and so verified. It is methodically objective when one is calculated by a verified method. Not all

---

[10] Besides the observations on point particles and reduction of the wave packet, mentioned above, long experience in experimentation have provided many examples of weirdness presented by conventional theory. Some relate to lack of definition of 'limits' in Heisenberg's Uncertainty that cause divergencies in track structure impact parameters [38] or in resolution of near field imaging in lithography *etc.* Other weirdness is solved by applications of equation 4 that are irrefutably verified by the practice of electron microscopy [18].

[11] Objectifying probabilities is in our view a short cut [cf. 1]. Probabilities are calculated.

probability amplitudes are verified; some are falsified. A false amplitude is like a false premise, it leads to no conclusion. Probability amplitudes that are neither verified nor methodically verified are not 'objective'. Not all amplitudes are measured: that would take too long[12]. We do not make elections over small decisions. To the credulous, axioms make-believe.

Returning to theory and logical truth, since the age of Locke, Newton, Hooke and Galileo, physics has been called empirical: observation trumped the oldest authorities. Popper best described the logic of science that is progressive, consistent, and falsifiable. The logic has impeccable epistemological pedigree. Those schismatics who claim that Heisenberg's Uncertainty Principles are axiomatic, or who by the same token claim measurement is instantaneous and discontinuous and essentially probabilistic, have discarded Popper's logic. What *we* find instead are three separate logical systems for three separate disciplines. The third of these is a dubious quantum theory, one that accepts axiom together with experimental fact for computational purposes. The result is not certain in the varied ways of math and physics; but is 'justified' in circumstances that require a best bet: a steeplechase with free reigns. It is an obvious conclusion that a 'computational proposition' is tested only by its capacity for creating useful numbers.

Our wave-group is a physical alternative: time dispersed and continuous. The conclusions of the three logical systems are not identical since physical explanation with internal motion substitutes for computational economy. What is possible and probable is a part of scientific effort, and that seems to be what quantum theory is. There is, moreover, a time for saying that the quantized wave-group is both quantum particle and wave: Born's wavefunction-as-probability-amplitude [51] corresponds to the complex wave-group in physics.

---

[12] However, as in all (Popperian) physical measurement that is in principle repeatable, a 'true' probability amplitude would be good for all time.

9. Conclusions

Condensation [2] is the process of resonant response, that occurs when a photon, or other wave-particle that is dispersed in time and space, is absorbed by or scattered from another particle that exists in a smaller space. Absorption can occur only if structural harmonies in the photonic probe and absorber are satisfied. The resonance occurs within the Uncertainty of the photonic wave-group by its internal motion. A resonance in anti-phase with the photon absorbed, causes photon annihilation.

This solution of a major, long-standing anomaly in Quantum Theory is derived from several features: wave-particle duality; the wave-group in relativity; Dispersion Dynamics; the definition of the quantum by the fact of dual harmonies proved necessary in quasicrystal diffraction; and the logic of physics:

Quantum mechanics (QM) is three disciplines with independent logical structures and methods:

1. Mathematical QM is axiomatic, consistent, certain—even of uncertainty. For example, the spherical harmonic quantum numbers in Schrödinger's solutions are of the greatest consequence for physics and chemistry. The wavefunction is complete only in the sense of the closed group of eigenstates.

2. Physical QM is explanatory, with meaningful propositions that are 'true' if not empirically falsified [10]. Given the wave-group, orthodox QT is now reconstructed; even condensation is explained as a general occurrence.

3. Computational QM is "crazy. It doesn't explain anything. It just gives you numbers"[16]. The *idea* of 'measurements' that are instantaneous in time—and therefore discontinuous—contradicts the way measurements are *experimentally* made. The 'objective probability amplitude' cannot be a physical object; it is only methodical. 'Quantum Theory' is a purposeful bet [13].

Only in physical quantum mechanics does the quantum have essence and existence: perceived 'whatness' and experimental reality.

The logical rules rouse fundamental questions: What is core to wave-particle duality? How does condensation occur? When we shine a laser beam onto a thin slit, the extended Fresnel image is transmitted—whether in near field or far field—whereas, when we

---

[13] Made with Einstein's unholy dice: "Subtle is the Lord, but malicious He is not" [13].

reduce the incident intensity with absorbers to the individual quantum rate of detection, a photon detector will detect quantum spots, typically on a fluorescent screen or photographic plate. When this experiment is repeated often enough, the multiple spots approximate to the laser image.

The same features occur with electron beam illumination as with the photons. Then, both the time uncertainty of the wave-group and the high velocity of the wave phase provide time and space *to enhance* resonant response. Molecules in the detection screen compete, by hidden variables (including phase coherence between probe and molecule, emission recoil *etc.*), for excitation; while probability theory predicts the excitations that are *methodically* measured.

These are not solutions to questions that mathematicians must ask. Mathematical beauty or computational beauty are fantasies when a feature is wrong or unexplained; but physicists have to offer an explanation for every type of event, and afterwards attempt falsification.

Dispersion dynamics, derived directly from Relativity, provides simple reasons for several misleading anomalies in QT: why the photon has positive momentum but no mass; why the electron has momentum proportional not to its wavelength but to the inverse of its wavelength; how its wave momentum is then shifted from the wavefunction real displacement into the complex plane; how, in special relativity, rest mass disengages the electron group velocity from its phase velocity; how the phase velocity is faster than the speed of light both theoretically and experimentally; the nature of intrinsic spin; how condensation of the wave-group is uniquely described by resonant response in the wave-group; and many other features. Physics abhors weirdness, whether quantum theory is understood as math, or physics, or both, or epistemically neither.

Appendix 1
Group velocity

Meanwhile, the concept of the group velocity, that is distinct from the phase velocity in equation 8, is illustrated by the beat frequency that is used by piano tuners. Consider two strings, one off-pitch. We can represent the real part of a carrier wave:

$$\text{Re}\{\exp(i(\omega t - kx))\}$$

(A.1)

That has maximum value when $(\omega t - kx) = 0$, Then:

$$v_p = \frac{x}{t} = \frac{\omega}{k} = f.\lambda$$

(A.2)

Superpose the second off-tune string, and sum the displacements:

$$\text{Re}\{\exp(i(\omega t - kx)) + \text{Re}\{\exp(i((\omega+\Delta\omega)t-(k+\Delta k)x)))\}$$

$$= \text{Re}\{\exp(i(\omega t - kx))\}.\{1+\text{Re}\{\exp(i(\Delta\omega t-\Delta kx)))\}$$

(A.3)

producing the *beat* envelope $\{1+ \text{Re}\{e^{i(\Delta\omega.t-\Delta k.x)}\}$ on the carrier equation A.1), with group velocity $v_g = \Delta\omega/\Delta k$. Finally, summing across the symmetric wave-group, gives:

$$v_g = d\omega/dk.$$

(A.4)

since the sum over a Gaussian distribution is Gaussian. The beat results in transport of momentum because of its phase difference from the carrier wave density.

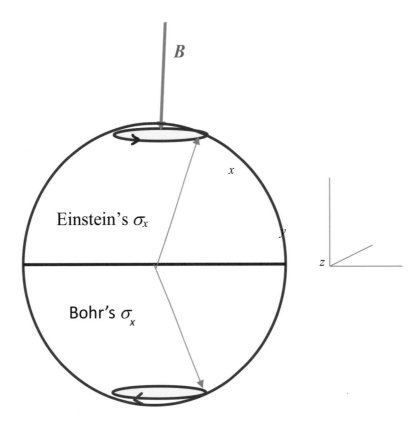

Figure A.2.1. Two indistinguishable electrons, disentangled from an $S=0$ spin state, conserve spin in the direction of an applied magnetic field. Not all components are measured: some are not determined. Orbital motion in phase space is illustrated in circles that contain directional arrows. If spin were not to be conserved, there would be is a serious failure in conservation laws, and that would be new physics.

Appendix 2
Postscript: EPR

The thought experiment of Einstein, Podolski and Rosen [15] remains a contention that divides the two opponents to the schism.

Notice that the interpretation of intrinsic spin as paramagnetism and diamagnetism in phase space [26] brings a new interpretation to quantum mechanics and Bell's inequalities. These are explained by Binney [52], where his final conclusion leads part of the way towards **figure A.2.1** in Dispersion Dynamics. However, the EPR thought experiment was about the commutator $[p_x,x]$ not $[\sigma,x]$. It is about entangled electrons, *i.e.* Fermions, not Bosons; Pauli's exclusion principle applies. The spin σ_x is a good quantum number on both Einstein's and Bohr's electron; but the other components σ_y and σ_z are undefined. So Einstein is right that conservation laws allow him to know Bohr's σ_x, but Bohr is right that he cannot know his σ_y and σ_z. Paramagnetism in phase space makes this debate irrelevant because the direction of the external magnetic field is easily controlled – but not conceived by Bohr.

Given the methods described in this treatise, wave-particle duality that is quantized by harmonies in the wave-packet, is analyzed by special relativity and dispersion dynamics. This leads to the representation of intrinsic spin as paramagnetic and diamagnetic resonance in phase space [26] due to applied magnetism to a charged particulate wave function. The two opposed orbital states are represented in the figure. Furthermore, in the absence of external applied magnetic field, para-magnetism in phase space is a response to orbital group motion and results in Russell-Saunders coupling of orbital and 'intrinsic' magnetic spins. This type of coupling is common in the lighter elements.

Among many theories for quantum mechanics, Bell has reviewed those of Bohr, von Neumann, Gleason, Jauch and Piron [52]. None of them account for the properties of the wave-group that are described here; theirs are mathematical theories by tone and design. They are surely successful in what they generically attempt, but by comparison are weakly explanatory and epistemically vague.

Table   A.3.1. Axioms of orthodox, computational quantum theory (in black, red and blue) compared with alternative open principles of empirical quantum physics (black, green and blue).

1. *Representational* ~~completeness~~ *of* ϕ.
The rays of **Hilbert space** correspond ~~one-to-one~~ with the physical states of the **wave-grouped** system

2. *Measurement.*
If the Hermitian operator $\bar{A}$ with spectral projectors $\{P_k\}$ is measured, the probability of outcome k is $<\phi|P_k|\phi>$.
These probabilities are ~~objective, i.e.~~
indeterminate **because of hidden variables**

3. *Unitary Evolution of isolated systems:*
   $|\phi> \rightarrow \mathbf{U}|\phi> = \exp(-\hbar^{-1}\mathbf{H}t|\phi>)$ is therefore
deterministic and continuous

4. *Evolution of systems undergoing measurement*
If Hermitian operator $\bar{A}$ with spectral projectors $\{P_k\}$ is measured and outcome k is obtained The physical state of the system changes
~~discontinuously~~: $|\phi> \rightarrow |\phi_k> = P_k\,(\phi)/\sqrt{(<\phi|P_k|\phi>)}$
continuously, in time $\Delta t \sim 8/\Delta\bar{\omega},$ **influenced by hidden variables**

Appendix 3.
Physical QM v Computational QT

A summary of the differences between the different logical systems of open physical quantum mechanics (green) and 'orthodox' computational theory (QT) (red) is given in **table A.3.1.** The differences are discussed at several points in the text

of this book. The representation of computational theory is adapted from blogs by Carroll. Physical quantum mechanics is explanatory; computational theory is a capricious number cruncher.

Moreover, physical quantum mechanics with dispersion dynamics is superior not only logically; but generally. Consider, for example, equations 5, 6,7 and 21 in the *non-relativistic* approximation $pc<<mc^2$: After substituting the Lorentz factor $\gamma = \sqrt{(1-v_g^2/c^2)^{-1}}$ and approximating

$$E \simeq m_oc^2 + \frac{m_ov_g^2}{2} = m_oc^2 + \frac{p^2}{2m_o}$$

(A.5)

that is the classical expression: mass energy plus kinetic energy.

Meanwhile, the *relativistic* expression is given by adapting de Broglie's hypothesis and equation 5:

$$p = m'\frac{d\omega}{dk} = m_o\gamma\frac{d\omega}{dk} = m_o\gamma c^2. \frac{\hbar}{\hbar}.\frac{k}{\omega} = \hbar k$$

(A.6)

where, from equation (7):

$$m_o = \hbar/c\sqrt{(\omega/c + k)(\omega/c - k)} \ ,$$

(A.7)

both particulate (first bracket) and wave-like (second bracket). By extension, these dynamics should apply to other elementary particles.

Reference

1. Bourdillon, A.J.,(2023) Quantum Mechanics: Harmonic wave-packets, localized by resonant response in dispersion dynamics, *Journal of Modern Physics*, (2022) **14** 171-182, doi: https://doi.org/10.4236/jmp.2023.142012

2. Penrose, R., (2023) Classical and quantum reality and the collapse of the wave function, *Youtube* https://www.youtube.com/watch?v=LKAphR6pBKQ

3. Frege, G., (1879), Begriffsschrift: eine der arithmetischen nachgebildete Formelsprache des reinen Denkens

4. Frege, G., (1893/1903 two volumes) Grundgesetze der Arithmetik,

5. Whitehead, A.N. and Russell, B (1910) Principia Mathematica, Cambridge University Press

6. Wittgenstein, L, ed. by Wright, G.H., Rhees, R. and Anscombe, G.E.M, Trans. by Anscombe, G.E.M., (1956) Remarks on the foundations of mathematics, Oxford

7. Nagel, E. and Newman, J.R. (1959) Godel's proof, Routledge & Keegan Paul

8. Bourdillon, A.J. (2012) Metric, myth and quasicrystals, UHRL, ISBN 978-0-9789-8393-2

9. James Clerk Maxwell, On Faraday's Lines of Force (1856).

10. Popper, K.R. (1980) The logic of scientific discovery Hutchinson, London

11. Popper, K.R., (1982) Quantum theory and the schism in physics, Hutchinson,

12. Popper, K.R., (1982) The open universe, an argument for indeterminism, Hutchinson

13. Pais A. 'Subtle is the Lord…' (1982) O.U.P.

14. Pais, A. (1981) *Niels Bohr's times in Physics, polity and philosophy* Clarendon

15. Einstein, A., Podolski,B. and Rosen, N. (1935) Can quantum-mechanical description of physical reality be considered complete? *Physical Review*, **47** 777-780

16. Feynman, R, (2016) Richard Feynman: Quantum mechanical view of reality2, *Youtube* https://www.youtube.com/watch?v=xNF_3KdpdrY

17. Susskind, L., (2016) Entanglement and complexity: gravity and quantum mechanics, *Youtube,* https://www.youtube.com/watch?v=9crggox5rbc

18. Bourdillon. A.J., (2023) Quantum Mechanics: internal motion in theory and experiment, *Journal of Modern Physics*, **14** [6] , 865-875. Special issue in quantum physics, doi: https://doi.org/10.4236/jmp.2023.146050

19. Bourdillon, A.J., (2013) A Travelling wavegroup II, antiparticles in a force field, *Journal of Modern Physics,* **4** 705-711, doi: 10.4236/jmp.2013.46097 https://www.scirp.org/journal/jmp .

20. Bourdillon,A.J., (2020) Relativistic approximations for quantization and harmony in the Schrodinger equation, and why mechanics is quantized, *Journal of Modern Physics,* **11**, 1926-1937 doi: https://doi.org/10.4236/jmp.2020.1112121

21. Bourdillon, A.J., (2022) Real quanta, with continuous reduction, *Journal of Modern Physics*, (2022) **13** [11] 1369-1381, doi; https://doi.org/10.4236/jmp.2022.1311085

22. Bourdillon, A.J., (2022) Harmony is cause—not consequence—of the quantum,   *Journal of Modern Physics*, **13** 918-931 doi: https://doi.org/10.4236/jmp.2022.136052

23. Bourdillon, A.J., (2015) A stable wave packet and uncertainty, *Journal of Modern   Physics* **6** [14] 2011-2020, doi: https://doi.org/10.4236/jmp.2015.614407

24. Bleaney, B.I., and Bleaney, B. (1965) Electricity and magnetism, Oxford

25. Bourdillon, A.J., (2012) A wave-group for entanglement, linking uncertainties in space and time, *Journal of Modern Physics,* **3** 290-296,
doi: https://dx.doi.org/10.4236/jmp.2012.330

26. Bourdillon, A.J. (2018) Dispersion Dynamical magnetic radius in intrinsic spin equals the Compton wavelength, *Journal of Modern Physics,* special issue on Magnetic Field and Magnetic Theory, **9** [13] 2295-2307
doi: https://doi.org/10.4236/jmp.2018.913145,

27. Dirac, P.A.M., (1958) *The principles of quantum mechanics, 4th ed.* (1958) Clarendon Press Oxford

28. Pais, A, Jacob, M., Olive, D.I., Atiyah, M.F. (1998) *Paul Dirac, the man and his work* Cambridge

29. Gell-Mann, M., (2008) Beauty and the truth in physics, *Youtube,*
https://www.youtube.com/watch?v=UuRxRGR3VpM

30. Belloc, H., (1907), John Henderson the unbeliever, *Cautionary tales for children,* Eveleigh Nash,

31. Jenkins, F.A., and White, H.E, (1937) Fundamentals of optics, McGraw-Hill

32. Hirsch, P., Howie, A., Nicholson, R.B., Pashley, D.W. and Whelan, M.J. (1977) *Electron Microscopy of Thin Crystals,* ch. 12.1 and appendix 4, *Krieger,* New York

33. Bourdillon, A.J., (2015) A Stable wave packet in the foundations of quantum mechanics, *Journal of Modern Physics* **6** [4] 463-471 (2015) doi; https://doi.org/10.4236/jmp.2015.64050

34. Bourdillon, A.J., (2017) *Dispersion Dynamics in the Hall effect and pair bonding in HiT$_c$,,* Nova science N.Y., ISBN: 978-1-53612-568-9

35. Einstein A., (2010) *Relativity: the special and general theory*, *Dover*

36. Bourdillon, A.J., Boothroyd, C.B., Kong, J.R., and Vladimirsky, Y., (2000) A Critical Condition in Fresnel Diffraction Used for Ultra-High Resolution Lithographic Printing, *J.Phys.D: Appl.Phys.* **33** 1-9

37. Bourdillon, A.J., and Vladimirsky, Y (2006) *X-ray Lithography – on the sweet spot* UHRL, CA

38. Bourdillon, A.J., (2000) Use of the Track Structure Approach in TEM, *Ultramicroscopy,* **83** 261-264 (2000) invited paper for commemorative issue in honour of Michael Stobbs

39. Bourdillon, A.J. and Stobbs, W.M., (1985) Elastic Scattering in EELS - Fundamental Corrections to Quantification, A.J. Bourdillon and W.M. Stobbs, *Ultramicroscopy* **l7** *147-150*

40. Aguilar-Benitez et al. (1992) Particle properties data booklet, Phys. Rev. **D45** part 2

41. Shechtman, D., Blech, I., Gratias, D. and Cahn, J.W. (1984), Metallic phase with long range order an no translational symmetry, *Physical Review Letters,* **53**, 1951-1953, doi:
https://doi.org/10.1103/PhysRevLett.53.1951

42. Bourdillon, A.J. (2021) Physical Quanta in Quasicrystal Diffraction, *Journal of Modern Physics*, **12** 1618-1632
doi: https://doi.org/10.4236/jmp.2021.1212096

43. Bourdillon, A.J., (2013) Icosahedral stereographic projections in three dimensions for use in dark field TEM, *Micron,* **51** 21-25,
doi: https://dx.doi.org/10.1016/j.micron.2013.06.004

44. Bursill, L.A., and Peng, J.L. (1985) Penrose tiling observed in a quasicrystal, *Nature* **316** 50-51

45 Kuhn, T.S., (1970) The structure of scientific revolutions, University of Chicago

46. Hare, R.M. (1982) Moral thinking: its levels, methods and point, OUP

47. Huntley, H.E., (1970) *The divine proportion* Dover

48. Bourdillon, A.J. (1987) Fine Line Structure Convergent Beam ElectronDiffraction in Icosahedral Al_6Mn, *Phil. Mag. Lett.* **55** 2l-26

49. Bourdillon, A.J. (2009) Nearly free-electron energy-bands in a logarithmically periodic solid, Sol.State Comm. **149** 1221-5

50. Bourdillon, A.J. (2010) Quasicrystal's 2D tiles in 3D superclusters UHRLbow LLC, ISBN 978-0-9789839-2-5, chapter 4.

51. Born, M., and Einstein, A., translated by Born, I. (1971) The Born-Einstein letters, Macmillan

52. Binney, J., (2011) Einstein-Podolski-Rosen experiment and Bell's inequality, ,*Youtube*, https://www.youtube.com/watch?v=uef_qN7VFuY

53. Bell, J.S., (1966) On the problem of hidden variables in quantum mechanics, *Rev. Mod. Phys.* **38** 447-452

Printed in the United States
by Baker & Taylor Publisher Services